International Conference on Ship Conversion, Repair & Maintenance 2021

Online
26 January 2021

ISBN: 978-1-7138-8304-3

Printed from e-media with permission by:

Curran Associates, Inc.
57 Morehouse Lane
Red Hook, NY 12571

CURRAN ASSOCIATES INC. proceedings.com

Some format issues inherent in the e-media version may also appear in this print version.

The Royal Institution of Naval Architects
8-9 Northumberland Street
London, WC2N 5DA
United Kingdom

Phone: 020 7235 4622
Fax: 020 7259 5912

publications@rina.org.uk

Additional copies of this publication are available from:

Curran Associates, Inc.
57 Morehouse Lane
Red Hook, NY 12571 USA
Phone: 845-758-0400
Fax: 845-758-2633
Email: curran@proceedings.com
Web: www.proceedings.com

International Conference

Ship Conversion, Repair & Maintenance

26 January 2021, Online Conference

PROCEEDINGS

THE ROYAL INSTITUTION OF NAVAL ARCHITECTS
8-9 Northumberland Street, London, WC2N 5DA, Telephone: +44 (0)20 7235 4622

The Royal Institution of
Naval Architects

INTERNATIONAL CONFERENCE

Ship Conversion, Repair & Maintenance

26 January 2021, Online Conference

Papers published in these conference Proceedings are peer reviewed.

THE ROYAL INSTITUTION OF NAVAL ARCHITECTS
8-9 Northumberland Street
London WC2N 5DA

Telephone: 020 7235 4622
Fax: 020 7259 5912

CONTENTS

FLOATING SHIPYARD: DRYDOCK DURING THE SHIP'S VOYAGE

A S I Ghowel, Senior Project Manager at Arab Shipbuilding and Repair Yard, Bahrain

SUMMARY

Shipyards are known as a land-based facility that ships steer to for docking and repair. This study represents a basic conceptual study for a new principal of developing a Floating shipyard: changing the phenomena of a fixed site shipyard into a self-propelled floating shipyard. This Floating Shipyard capable to travel and conduct drydock activities at the client's location or even can lift the client's vessel and travel to its next designated location, taking advantage of completing the dry dock and repair during the voyage. The arising challenges, to this floating shipyard, such as lay-offs, restructuring, and environment legislation requires a balanced solution. This solution, of all this, relays on the adaptation of talent management and competitive production tools, during this huge undergoing changes in its backbone structure. The Floating shipyard, here named "F-Yard", changes the mindset of dry-dock. This is because F-Yard travels to the client's location or capable of pick the client's ship towards the cargo destination, where it can complete the required drydock or repair. The cutthroat advantage of the F-Yard comes from self-propelled where the other approaches depend on others for mobilizing and anchoring from one location to another. Plus, F-Yard could serve other industries, due to its fully equipped workshops, that able to serve different market needs such as oil and gas and renewable energy...etc. These open the door for a variety of business scope. F-Yard depends on front end engineering and marketing teams to optimize its route and to sort the supply chain requirements.

1. INTRODUCTION

The idea of the "F-Yard", changes the mindset of dry-dock. This is because F-Yard travels to the client's location or capable of pick the client's ship towards the cargo destination, where it can complete the required drydock or repair.

In another way, it targets the customers that are missing any shipyard in their working area/ route or the ones that find the shipyard is not matching their needs from any perspective, such as quality, safety, time frame…etc., which causes extra time and money due to the off-hire period from the charter.

The cutthroat advantage of the F-Yard comes from self-propelled where the other approaches depend on others for mobilizing and anchoring from one location to another.

Plus, F-Yard could serve other industries, due to its fully equipped workshops, that able to serve different market needs such as oil and gas and renewable energy…etc. These open the door for a variety of business scope.

F-Yard depends on front end engineering and marketing teams to optimize its route and to sort the supply chain requirements.

2. "NO SHIPYARD MARKET"

Author believes that there is a new coming market named" No Shipyard market", where the major companies try to present different solutions to suit the areas where there are no, or poor shipyards exists.

For example, DORMAC Shipyard presents complete mobile operational offshore old platform, referred to as "Shipyard-in-a-Box", with a full spectrum of site facilities that can lay on a berth for a long time and work as a workshops to serve the surrounding ship, but without a docking option, only in afloat condition.

Another example, by Boskalis through Dockwise Vanguard, that's able to carry cargoes up to 110,000 tones. She is used for dry-docking of FPSO while operation.

Also, there is Suez floating dock with built-in workshops and overhead crane to cover the vessels that are passing in Suez Canal.

Other is a heavy lift ship to serve as marine Maintenance and Repair [MAR] platforms. This study was prepared by Robert Cullen [2007]

These solutions attract the ship's owners, not only for its mobilizing possibility, but it exceeds this as while coming to the ship's location can source the required material, for repair, in the way allowing the preparation and prefabrication to start.

Moreover, the floating shipyard acts as transport for the required spares for the dry-dock instead of hiring 3rd part carrier to get these spares to the shipyard's location.

Another windfall gain for the client-side is that is ship resources oriented literally because all the yard resources are put together to complete the project in a short duration. This means there is no priority list or emergency project or even a shortage of man-power due to holidays.

The main customer is the offshore units, for instance, any Jack up will save huge wasted time and money for mobilizing and demobilizing to and from the nearest shipyard with a limited speed.

Ultimately F-Yard works as a floating dock, thus it can be chartered to any shipyard during the peak market.

3. MANAGEMENT METHODOLOGY

The operation method for F-Yard will be a merge between Agility and Lean founded on Industrial 4.0.

This allows us to have an accurate, valuable, and fast response. Plus, facing the unclear, changeable, and interlinked scope of any emergency or normal DD.

The more advantage of the adaptation of this merge is the cost reduction for the shipyard expenses and DD cost. Moura (2011) described it as a "high level of rapid reconfiguration and will eliminate as much waste as possible".

The Key factor for applying agility lean manufacture is the Centre Planning Unit- CPU. At this stage, the flexibility of the F-Yard must be notable and credible than others.

Lamp [1986] "Engineering for Production" explored the Design for Production (DFP) approach, that is used for the correlation of production design with the available or planned facilities and production methods.

This means, in case of an emergency or a normal DD, it starts by the CPU which designs the full production process and utilizing different simulation scenarios, to achieve the best layout for optimizing the production flow. In respect to a risk assessment study and TPS wastes.

Also, in the project management level, the reason behind selecting APM methodology, for each DD project was explained by Daniel [2009] "Agility is the ability to act proactively in a dynamic, unpredictable and continuously changing environment.

And, illustrated more by Owen et al, [2006] that, "an organizational agility is the ability to be inherently adaptable to changing conditions without having to change."

In other worlds, responding to the dynamic, changeable and uncertainty environment of DD projects, there should be a controllable, stable and fast approach like agility.

Chin (2004) gives an equation for Agile PM Environment = [Uncertainty + Unique Expertise] x Speed.

Back to Daniel, he listed down the advantages of the agility approach, which fulfilled F-Yard needs, as follows:

1- Team members collaborate in all aspects: Shipyard staff should be a high competence, multifunction expert team to be allocated in different areas.

2- Regulation of changes depends on flexible and adaptable procedures: as there is no certain scope.

3- Customer is continuously involved throughout the project lifecycle: as he is living onboard F-Yard, he is part of day to day activity.

4- Testing is well defined and done more frequently: testing at each stage is mandatory for the readiness final trials without having cost of quality.

5- Technology: by implementing pro-active innovation strategies.

Utilizing different tools to increase the exploitation of the working area and eliminating the production wasted, such as Systematic layout planning (SLP) and Analytic Hierarchy Process (AHP), presented by Matulja [2015].

Therefore, literally, for each DD project there will be a separate workshop layout and a production plan with respect to the designated area for necessary material storage.

This flips the space restriction of F-Yard into a markable advantage. Thus, this adding value translated to the customer by achieving better Takt Time.

Part of the operation plan will be implementing the Industrial 4.0 system, for the F-yard machinery. This is very essential not only to allow communication between one equipment to another but also to signpost if any unscheduled maintenance required or expected failure for the F-Yard's equipment.

This will ensure the continuity and reliability of all the machines during a short period of drydock in the middle of the water.

The Workshops in F-Yard are based on ISO Containers: that represents a fully equipped workshops of the different trades. This principal gives F-Yard the ability for having different arrangement/layout based on the coming DD requirements and optimizing the limited space of the F-yard.as shown in Figure [1].

This is the key factor for eliminating the space constrain and to implement the lean production.

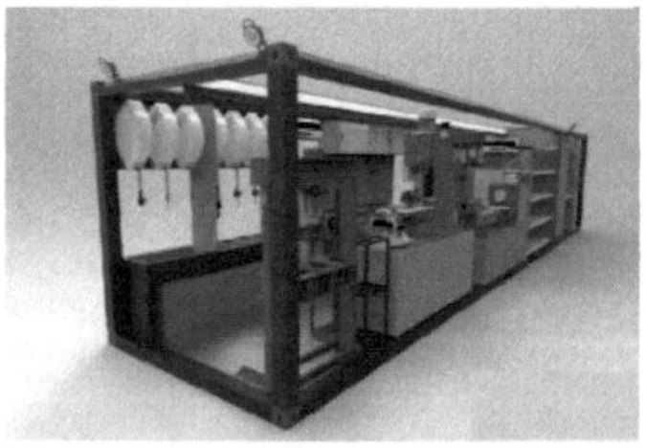

Figure [1]

4. OPERATION PARTICULARS

The general design outcome is approx. 240 mL by 40mB to be able to accommodate an offshore unit (alongside) and to dock a drillship type. Figure [2] shows a draft model to F-Yard.

Looking at the bigger picture, this F-Yard will be capable to work as a floating workshop serving different industrials creating more variety of business scope, and the outline of the weather conditions fort this was determined as Max Beaufort force 2 [4-6 Kts Wind - 0.2 m wave height] .

Also, it will be capable to serve ship inside the dock with 60,000 DWT plus other two unites [ships/offshore] alongside same time.

There are two proposed general arrangement for the F-Yard, one if open and other is closed. This to fit the purpose for the end user.

5. Layout and Main Components

The main components are listed down in Table [1].

Feature	Requirements
Heavy Lift	Vessel for Lifting vessel up to 60,000 DWT, with open deck from both sides.
Rail	on both sides, for floodable 50-ton SWL crane shifting and automated Robert preservation equipment.
LMCS	Floodable Light Modular Causeway Sections, that is capable of lock the containers in any position.
Workshop containers	Shelter space in form of 20 and 40 ft ISO, works as fully equipped containers.
Multifunction Tools	forklift with a man basket and container handler
Helipad	One of the SWOT outcome to mitigate and avoid emergency cases.

Table [1]

Figure [2]

6. Business Proposal

The floating shipyard idea proves to present an inimitable solution by the self-propelled heavy lift that has the proficiencies to perform all the docking activities of the routine or emergency docking, in a station location or while voyage from one location to another.

This concept challenges new market "No Shipyard Market" sector which is looking to minimize or eliminate the off-hire time due to traveling to the shipyard and the time consumed for dry-dock. All this time can be utilized while the Floating shipyard is coming to the ship's location and doing the work in place.

Adding to the ship repair, the main revenue expected to come from the offshore units i.e. Jack up, drill ships…etc. As serving it in its operation place saves huge wasted time and money to and from the nearest shipyard with a limited speed.

This comes from identifying the wasted time and cost that the client is suffering from during the necessitate drydock to maintain the Class certificates.

Looking to the above advantages will magnetize the ship's owners to this idea as the cost comparison between hiring this floating shipyard is much less comparing to the amount that is lost in the drydock and to and from the drydock as well.

Means of the cost-saving will be in the ship's owner side by a simple calculation. Moreover, the floating shipyard can act as transport for the required spares for the dry-dock instead of hiring 3rd part carrier to get these spares to the shipyard location.

In accordance with the above, this floating shipyard serves other markets as well. This comes from the basic principle design that any shipyard can accommodate other business scopes, with few amendments, such as the new building, oil and gas sector or renewable sector…etc. Plus, the its capability as heavy-lift transport from one location to another.

Another windfall gain for the client is that is ship resources oriented literally because all the yard resources are put together to complete the project in the shortest duration. This means there is no priority list or emergency project or even a shortage of man-power due to holidays.

Finally, F-Yard works as a normal ship means it can be chartered to another shipyard during the peak market to cover the load coming to one shipyard.

6.1 RISK ASSESSMENT

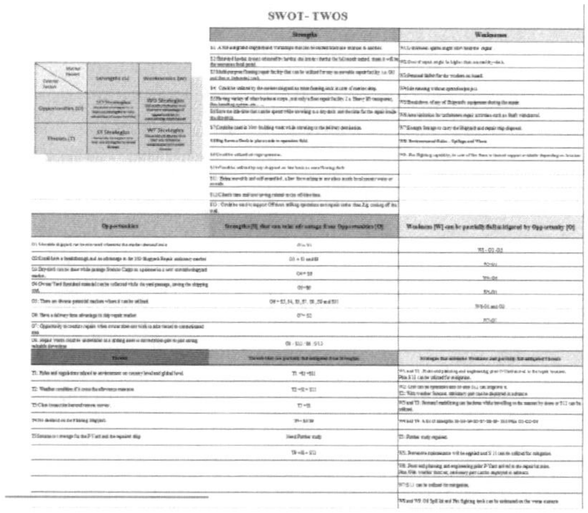
SWOT- TWOS

Figure [3]

Inside this comprehensive study as shown in Figure [3] , there is detail expansion for the SWOT and TOWS analysis along with other risk assessment tools and main outcome of this are the below findings:

• In conclusion to the above SWOT and TWOS analysis, it is clearly shown that the idea is based on re-construction the backbone of a traditional shipyard into a self-propelled floating shipyard. This new Floating Shipyard talented of repositioning itself based on the market opportunity and demands, this shows a significant revenue and profit due to different potential business list.

• Top of this list is the multi-purpose business scope, which means it is not limited to a shipyard activity or even shipping industrial, but it targets diverse industrial markets, such as sustainable energy and heavy lift.

• Back to the Shipyard renovation, the main advantage of the Floating shipyard that it is capable to eliminate the off-hire time that the ship's owner losing while traveling to and repair inside a physical shipyard for repair along with other ships and rigs and moreover, it can take advantage by doing the dry-dock while traveling from/to the Ship's cargo. what is more, as the floating shipyard will be resources oriented to one project, this gives the plus of early completion with the highest quality and reliability.

• Further to the investigation above, these analyses prove that the challenges that face the idea could be mitigated or eliminated by permanent solutions. Above all, the one related to the environment and supply chain, which can be mitigated by utilizing the main crucial benefit of the repositioning capabilities as floating self-propelled shipyard.

• With respect to environment concerns, green generates from using different processes and systems that has an environmental impact for both energy use and pollution of null.

• Bottom line, in the rising of the new NO shipyard market, Floating shipyard has the prospective to lead this market purely in favor of its self-mutable capabilities, and moreover, it can share other markets as well to give surety its revenue and profit.

• There is quite challenge for the arranging of the spare parts that can be mitigated and shared by the planning and modelling for the up coming projects.

7. FINANCIAL STUDY

Financial study has been done with business outlook represents high PV and IRR. This study was based on a primary market scan for the previous mentioned case study and SWOT and TOWS analysis.

All references that are used to build this model are used to show the potential of the selected region with respect to customer needs and market demands towards this new idea.

7.1 CASE STUDY

This market study was based on a realistic case study for a location "West Africa", considering the shipyards that are in this region as competitors plus the amount of the vessels that are visiting this area.

Selected country, in this area, is Gabon. This area was selected based on the oil exploration and drilling activities that are rising recently and this country selected based on its geographical location that serves Centre and West Africa plus its top rank with respect to security level compared to the surrounding countries.

7.2 MARKET NEEDS

Table [2] shows the shipyards that are in this region with respect to their facility of graving and floating docks.

Facility	*QTY*	*Length[m]*	*Breadth [m]*
Graving Dock	1	277.4	45.7
Graving Dock	1	191	25
Graving Dock	1	106.7	13.7
Floating Dock	1	235	38
Floating Dock	1	210	36
Floating Dock	1	172	33
Floating Dock	1	140	32
Floating Dock	1	122	28
Floating Dock	1	104.8	23
Floating Dock	1	90	25.6

Floating Dock	1	80	16
Floating Dock	1	60	13.5
Floating Dock	1	42	13.5

Table [2]

And Chart [1] shows the number of the vessels calling this area with respect to their type.

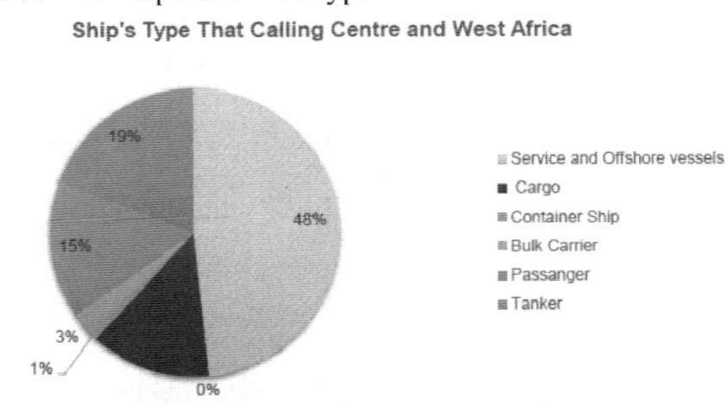

Chart [1]

7.3 FESABILITY GLANCE

Based on this segregation, calculation has been estimated for small to medium size project based on realistic data from drydock data base.

Another important profitable factor is the daily rate for each type of vessel. This will be considered as ship's owner gain for avoiding the off hire by going to a shipyard.

Sametime, F-Yard will share with a parentage this saving for its service and mobilization

Table [3] shows the number of offshore units that are working in this area.

IHS Markit

International Rig Market Analysis by Region			
	WEST AFRICA		
	SS	DS	JU
WORKING	1	8	17
IDLE	3	15	7
UTIL %	25.0	34.8	70.8

Table [3]

Table [4] shows the daily rate for the Jack-up and Floater based on March 2020 rates:

Curr. dayrate estimates ($USD):

Category	Dayrate
Premium 400ft IC jackup	92 000
Vintage jackup (300ft+ IC)	65 000
Premium HE jackup (ex CJ70)	115 000
Vintage HE jackup	85 000

Curr. dayrate estimates ($USD):

Category	Dayrate
6th/7th gen drillship	225 000
Vintage drillship	115 000
6th gen semisub	185 000
Vintage semisub	120 000
6th gen HE semisub	325 000
Vintage HE semisub	190 000

Bassoe Analytics

Table [4]

Offshore vessels daily rates as Chart[2] .

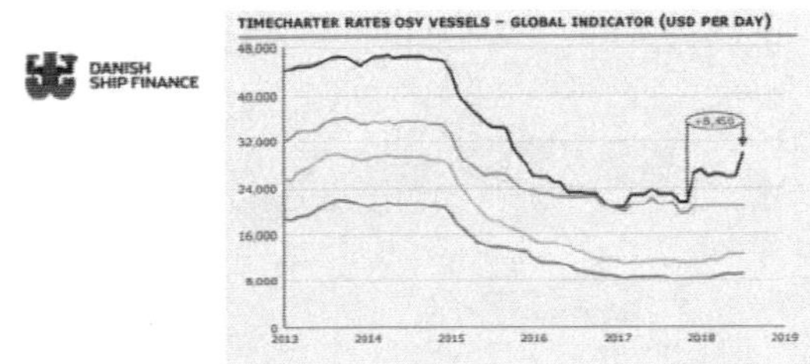

Chart[2]

Finally, the calculation was done for an interval of 5 years from 2012 to 2016, with respect to the inflation factor. Based on this Table [5] shows the expected occupation details and Chart [3] shows the related curve.

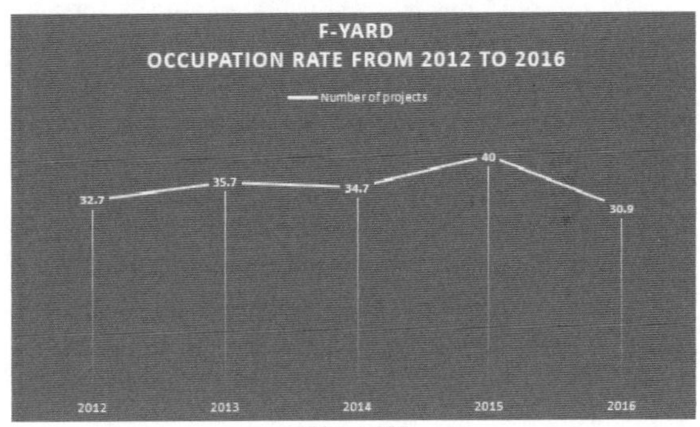

Chart [3]

18 vessel + 1 Rig	18 vessel + 2 Rig	19 vessel + 1 Rig	21 vessel + 2 Rig	17 vessel + 1 Rig
Year 1	Year 2	Year 3	Year 4	Year 5

Table [5]

This with respect of all types of costs: direct, indirect, fixed and variable. The outcome, in terms of total cost as per Chart [4] and related ROI per Chart[5].

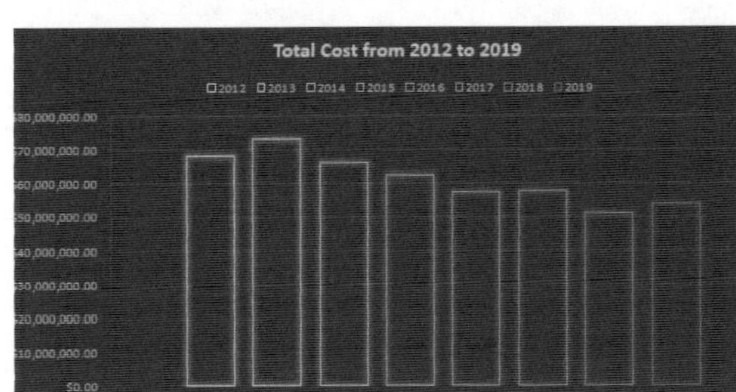

Chart [4]

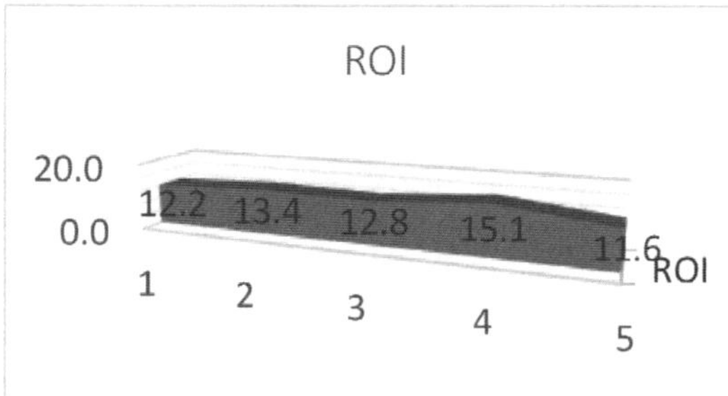

Chart [5]

8. RESOURCES PLAN

The primary assumptions about manpower requirements, as Table [5]. In addition, for the required list of equipment that need for each container and major equipment for the shipyard itself, all have been listed down in the comprehensive study. to estimate when cash will begin to flow into the F-Yard.

Job Function	Nr. Per Shift
Project Manager	1
Quality Manager	1
Purchaser	1
Total	3
Yard Personnel	
Engine Fitter	4
Engine and Steel Foreman	1
Steel Worker	3
Pipe Fitter	2
Cranage	2
Electrician	1
Insulation	1
Painting foreman	1
Cleaning, Scaffolding and Painting	5
Total	20
Grand Total [day/annual]	43/66
Indirect labour for Safe manning for vessel	Amount needed
Captain and Dock Master	1
Mate and HSE Manager	2
Sailors	2
Engineers	2
Kitchen and Housekeeping	4
Grand Total	13

Table [5]

9. FURTHER STUDY

Detail study has been done as first stage to initiate the feasibility study of this turnover project which conclude the feasibility of the project and the profitability of it.

Further study is required to be done when the project sponsored by company for implementation.

10. CONCLUSIONS

Figure [5]

Time is an essential factor in our shipping industrial, therefore a new approach for the docking the vessels needs to be implemented. This can be done by changing the backbone structure from a normally fixed-site shipyard to a Floating Shipyard.

For this unique solution, there is a lot of homework that needs to be done to tailor the industrial tools and management approaches to suit this solution requirements.

F-Yard, Figure [5] fast flexibility and inclusive components and allow it to be a multi-purpose floating workshop that can be utilized in dissimilar industrials.

Finally, obstructions are made to be crossed by thinking without the box, where the author believes that boxes are the land-based shipyard.

11. REFERENCES

Mossman, Alan (2009) Creating value: enough way to eliminate waste in lean design and lean production. *Lean Construction Journal* 2009 pp 13 – 23Bahadir

Dcrust Murthal, Sonepat and Haryana, 2013. An Introduction to Applicability of lean in shipbuilding. *International Journal of Latest Research in Science and*

Technology, [e-journal] 2(6). Available at www.mnkjournals.com.[Accessed 2 April 2015]

Herrala, M.E., Pekuri, A and Aapaoja, A. 2012, How Do you understand Lean*? 20th International Group of Lean Constructions*, San Diego, USA, 10 Oct 2012, USA, IGLC.

Philip c. Koenig, Hitoshi Narlta and Koichi Baba, 2002. Lean Production in the Japanese Shipbuilding Industry? *Journal of Ship Production*, 18(3), pp.167-174.

Richard Lee Storch ,1999, Improving flow to achieve lean manufacturing in shipbuilding, *Production Planning & Control: The Management of Operations*, 10(2), pp. 127-137

Thomas C. Rudzewicz, Jonathan H. Potterton, Katherine M. Lappe, Brent C. Yezefski and David J. Singer, 2010. A new Approach to ship repair Using CONWIP and Parts Kits. *Journal of ship production and design*, 26(2), pp. 155-162.

Wirahadikusumah,R.D and Sulistyaningsih D., 2013. The role of owners in the supply chains of highway construction projects: An overview of Indonesian cases. *21st International Group of Lean Constructions*, Fortaleza, Brazil, 6 Sept 2013, USA, IGLC.

Womack, J. P., Jones, D.T., and Roos, D., 1990. The Machine that Changed the W world. New York: Rawson Associates.

Izunildo Fernandes Cabral. (2011) An Information Model for Lean, Agile, Resilient and Green Supply Chain Management. Master. Thesis, The Faculdade de Ciências e Technologic da Universidade.

Daniel Adjei. (2009) Application of Traditional and Agile Project Management in Consulting Firms. Master. Thesis, Umeå University.

Daniel I. Janson. (2016) THE DEVELOPMENT OF A GREEN SHIPYARD CONCEPT. Master. Thesis, University of Twente.

Martijn .L.D.Schuttevaer. (2013) Boskalis offers combined refloat and dry dock solution through SMIT and Dockwise

Thomas Lamb. (1986) *Engineering for ship production.* Michigan: The university of Michigan.

Robert Cullen. (2007) *Use of Heavy Lift Ship as a Maintenance and Repair Vessel* (MD 20817-5700). Boulevard west Bethesda: Naval Surface Warfare Center.

Mark V. Arena et al, (2004) *Monitoring the Progress of Shipbuilding Programmes* (online). Europe: RAND Corporation. (Accessed 2005).

Mr. William G. Becker. (1993*) IDENTIFICATION OF NON-VALUE-ADDED TASKS IN SHIPBUILDING*

Ahmed Samir Ghowel. (2015) Customer's Waste at Dry-docking. In: SNAME, *world Maritime Technology Conference, 3* November. Rhode Island Convention Center and Omni Hotel.: SNAME.

Onno A.J. Peters. (2013) MCE Deepwater Development. In: The Hague, The Netherlands. *Changing the Heavy Marine Transport Market,* March. Dockwise: The Netherlands.

T. Terpstra, E.A. Hellinga and H.C. Leerdam. (2013) Offshore Technology Conference. In: OTC program committee, *Offshore dry-docking of FPSOs; a response to industry needs,* 29–31 *October*. Rio de Janeiro, Brazil: OTC.

D.A. Moura and R.C. Botter (2011) Sustainable Maritime Transportation and Exploitation of Sea Resources. In: Taylor & Francis Group, London, *Can a shipyard work towards lean shipbuilding or agile manufacturing.* London, September. London: ISBN.

12. AUTHORS BIOGRAPHY

Ahmed Samir Ghowel is Senior Project Manager in Arab Shipbuilding and Repair Yard- ASRY. He has been working in dry-docks, conversion and repair projects for almost thirteen years in different shipyards.

He received his BSc from Alexandria University in Naval Architecture and Marine Engineering. Thereafter, he received his MSc from Liverpool University in Construction Project Management. He was registered as Incorporated Engineer from Engineering Council-UK.

His research interests are: Lean production system and Six Sigma in Shipyard. He conducted different lecturers and presentations in different locations. He was awarded by the US Navy for his management approach in handling major repair for US Navy vessel.

REMOTE INSPECTION OF A FPSO UNIT CRUDE OIL TANK WHILST STILL PRODUCING

A Comley, A Comley Limited, UK
C Ross, Bumi Armada UK Limited, UK
S Moir, Texo DSI

SUMMARY

In 2020, with the unit still producing a crude oil storage tank on the Bumi Armada owned and operated Armada Kraken FPSO was successfully inspected by Texo using a UAVS flown from outside the tank. This type of truly remote survey significantly reduces the risk to personnel, the survey time and the survey cost, whilst maintaining a survey standard that is close to that achieved by personnel entering the tank and utilising scaffolding.

Lessons learnt in the following areas are discussed:
- Detailed Scope of Work value, including potential & actual anomaly handling.
- Careful pre-mobilisation briefing of the project team and Class surveyor.
- Engagement with Bumi Armada personnel, onshore and offshore, and the Texo project team.
- Engagement with DNVGL to enable the inspection to be Verification & Class credited.
- Development by Texo of project specific procedures based on onshore trials.
- Tank cleanliness and cleaning.

NOMENCLATURE

ABS	American Bureau of Shipping
APL	Advanced Production & Loading AS
BAUK	Bumi Armada UK Limited
CAA	Civil Aviation Authority
CAP	Civil Aviation Publication
ClassNK	Nippon Kaiji Kyokai (Japanese Register of Shipping)
COT	Crude Oil Tank
CVI	Close Visual Inspection
DNVGL	Det Norske Veritas Germanischer Lloyd
DVI	Detailed Visual Inspection
FPSO	Floating Production Storage & Offloading unit
GVI	General Visual Inspection
HIMS	Hull Integrity Management System
HSRBIP	Hull Structural Risk Based Inspection Program
IACS	International Association of Classification Societies
IMDG	International Maritime Dangerous Goods Code
IVB	Independent Verification Body
LR	Lloyd's Register
PMAs	Planned Maintenance Activities
NOV	National Oilwell Varco
OIM	Offshore Installation Manager
RBI	Risk Based Inspection
ROV	Remotely Operated Vehicle
R&LE	Repair and Life Extension
SIMS	Structural Integrity Management System
SIRM	Structural Inspection Reporting Manual
STP Buoy	Submerged Turret Production Buoy
TA	Technical Authority
TSIMS	Topsides Structural Integrity Management System
UAV	Unmanned Aerial Vehicle, colloquially known as a drone
UAVS	Unmanned Aerial Vehicle System (UAV + Payload + Control System)
UK	United Kingdom
UKCS	United Kingdom Continental Shelf
WBT	Water Ballast Tank

1. INTRODUCTION

The Armada Kraken Floating Production Storage FPSO unit, which is owned and operated by Bumi Armada, is moored on the Kraken Field, in the Northern North Sea of the UKCS, see Figure 1, on behalf of EnQuest and partners. The FPSO is based upon the conversion of the 2007 Hyundai Heavy Ind. Co. Ltd built Suezmax sized conventional trading tanker Prisco Alcor.

The tanker was classed from new with DNVGL, with the notation 1A1 ICE-1A Tanker for Oil with additional notations ESP SPM E0 VCS-2 CLEAN TMON NAUTICUS (new building). Between March and April 2014, the vessel underwent a second classification renewal survey prior to the commencement of the repair and life extension (R&LE) project at the Keppel Offshore & Marine shipyard, Singapore. The aim of the project was three-fold, namely:
- To repair any existing damage;
- To convert the vessel for use as a FPSO, and;
- To strengthen the hull as to withstand the expected operational conditions in the North Sea.

As the subject of this paper is the in-service inspection of a Crude Oil Tank (COT), the conversion itself is not discussed, except where it has an influence on the inspection regime. Post-conversion, DNVGL classed the Armada Kraken FPSO with the main notation +OI Ship Shaped Oil Production Storage Installation with additional notations POSMOOR, HELDK, CRANE, and BIS.

The Armada Kraken FPSO was subsequently towed from Singapore to Rotterdam, arriving in January 2017, for final inspections and trials. The FPSO was then towed out and hooked up to the previously installed STP Buoy in February 2017, with first oil being produced on the 23rd June 2017.

2. GENERAL ARRANGEMENT

The MT Prisco Alcor was a typical double-hulled Suezmax tanker, with six port & starboard crude oil tanks (COTs) and water ballast tanks (WBTs) numbered from forward aft, plus fore and aft-peak WBTs, and port & starboard slop tanks just aft of number 6 COTs. The accommodation and machinery space being aft, with a pump room just forward of the machinery space adjacent to the slop tank, as outlined in Figures 2 & 3.

As the Armada Kraken FPSO is moored via a twelve-line mooring system, consisting of three clusters of four lines each, connected to a freely weather-vanning turret, the port & starboard number 2 COTS had a moon-pool for the turret installed on the centreline and were subdivided into a forward pump room, production water tanks and bulk storage tanks all associated with the operation of the wells and their downhole hydraulic submersible pumps. Thus, reducing the COTs from 6 pairs to 5 pairs, but also introducing a number of additional tanks that also have to be inspected over time and included in the Structural Integrity Management System (SIMS).

Additionally, a series of modules, as shown in Figure 2, ranging from around 170 tonnes to more than 2,225 tonnes were installed on stools that were integrated into the ship structure through the main deck. The following were also integrated into the hull structure during the conversion:

- Forward flare tower;
- Turret deck house, complete with pull in winch and pedestal crane;
- Elevated forward bulwark for green water protection;
- Aft cantilevered helideck;
- Aft offloading hose reel;
- Aft hawser handling winch, and;
- Port & starboard pedestal cranes.

3. INSPECTION REGIME

3.1 STRUCTURAL INTEGRITY MANAGEMENT SYSTEM

During the conversion and R&LE of the Armada Kraken FPSO, a Structural Integrity Management System (SIMS) based on a Risk Based Inspection (RBI) process was developed. This in essence consisted of a high-level SIMS, which incorporated a Hull Integrity Management System (HIMS) and a Topsides Structural Integrity Management System (TSIMS). As this was undertaken prior to commencement of field operations and initial operational surveys, then whilst the HIMS provided a good basis for inspections, it was agreed between BAUK, DNVGL and the independent Marine Technical Authority (Marine TA) for the asset that it could be improved upon. Consequently, a revised HIMS was developed for the Armada Kraken FPSO that could be dovetailed into the existing HIMS and so SIMS. This needed to satisfy both the UKCS offshore oil & gas safety statutory legislation requirements in terms of Safety Case and Verification activities, and DNVGL Class requirements. The revised HSRBIP was subsequently approved by DNVGL, acting as both the Independent Verification Body (IVB) for the asset under the UK Safety Case regime and as Class.

For simplicity and practicality, the boundary between the HIMS and TSIMS is defined as the top of the stools that support the topside modules. This boundary was chosen as it's a clear demarcation point during surveys and was also the boundary used for the Structural Finite Element Analyses for the hull and topsides during the conversion process.

3.2 INSPECTION PLANNING & DEFINITION

The Hull Structural RBI Program (HSRBIP) has inspection intervals of between 9 and 13 years for the main COT structures and pipework, however, as the tanks were all previously inspected during the conversion process then this means that their inspections all become due at the same time. This is obviously convenient for a trading tanker that goes into dry dock every five years but is less than convenient for a floating production unit that is designed to stay on station for more than twenty-five years, as all the COTs would need to be taken out of service and inspected during a single year, which would be impractical from a production and operational perspective. It is also a long time to go on a recently converted hull without gathering any condition data, particularly regarding any issues that may have been inadvertently created during the conversion process. Consequently, it was agreed between the BAUK Marine Superintendent and the independent Marine TA to instigate a rolling tank inspection program, with the aim of always staying ahead of the HSRBIP, whilst also attempting to minimise the operational impact of the inspections. At the same time, it was agreed that wherever possible remote inspection techniques should be used to minimise the risk to personnel carrying out the inspections, but without compromising the quality of the inspections.

To enact this methodology within the Planned Maintenance System, instead of having Planned

Maintenance Activities (PMAs) for inspecting each of the tanks and elements within them, the system has been simplified to the following PMAs:

- RBI Lookahead to identify scopes of work for the following year for all hull structural inspections;
- RBI inspection of subsea exterior hull and openings;
- RBI inspection of above water exterior hull, including accommodation block, exhaust stack, helideck, offloading reel base, crane pedestals, flare tower;
- RBI inspection of Crude Oil Tanks (COTs);
- RBI inspection of Water Ballast Tanks (WBTs);
- RBI inspection of internal spaces, integral tanks not covered elsewhere, support stools for process modules, Deck,
- RBI inspection of Turret
- Inspection of Cathodic Protection System, and;
- Inspection of Exterior Cathodic Protection System.

Where the groupings take into account inspection synergies, for instance the subsea exterior hull inspections are undertaken via Remotely Operated Vehicle (ROV) whilst the above water exterior hull inspections are undertaken using an Unmanned Aerial Vehicle (UAV), colloquially known as a drone, which can at the same time inspect the helideck structure, flare tower, crane pedestals and other areas that would otherwise require rope access techniques and the inherent risks associated with using them.

Based on the experiences of the BAUK Marine Superintendent and the independent Marine TA, it was decided early on that BAUK should aim to keep control of all inspections. This might seem rather obvious, a company wanting to keep control of a survey, but in practical terms survey specifications can often be little more than an instruction to a survey company to inspect tank X or the exterior hull of the vessel, without any real regard for what level of survey is required. The net result of poorly specified surveys is that it can be difficult to establish whether the survey has been completed successfully, whether it has been value for money and whether it is providing real integrity assurance or simply an illusion of it. Additionally, it can be difficult to compare survey results from two different companies undertaken some time apart and to work out whether a reported anomaly really is an anomaly, and if it is not then going through the process of persuading Class that it is not when the Class approved inspection company has said that it is. Hence being proactive in this area and developing detailed inspection work-scopes, particularly where ROVs or UAVs are concerned, has real benefits.

The inspection scopes of work developed by BAUK have been put together in such a way as the information is there for all the COTs for instance, and only a simple table up front needs to be changed to specify the tanks for that year. This has several benefits, the first and most obvious being that once developed it only need to be reviewed and tweaked as necessary each year thus presenting a substantial cost saving and ensuring consistency in specification. Secondly it means that if the specified tanks can not be inspected for some reason, but others are available then the inspection can be switched to the available tanks rather than cancelled, thus ensuring a continuous survey regime as desired by BAUK. It also means, given that the inspection scopes of work all contain inspection worksheets that must be filled in during the inspection, that all items requiring inspection are systematically surveyed, and that the results from surveys a number of years apart can be easily compared as the reporting follows the same format each time.

Additionally, based on the issues experienced where inspectors have mis-reported something as an anomaly, the classic example on a floating asset being the mooring chain corroding when that is expected and the chain size contains a corrosion allowance, the inspection scopes of work also make use of the following three containers:

- Anomaly – something that is clearly cracked, broken damaged or missing;
- Item of interest – something that might be an anomaly but is not clear cut;
- Inspection Note – something that has previously been reported as an Item of Interest or an Anomaly that poses no risk unless excessive e.g. mooring chain corrosion.

By adopting these containers and having reporting systems associated with them, the misreporting of anomalies is generally avoided, particularly when combined with a thorough pre-mobilisation briefing of the inspection team and Class surveyor. The latter to make sure they understand the intended extent of the survey, the inspection methods being employed, any special instructions i.e. such as do not speak to the UAV pilot whilst they are flying (as discussed later), and the item of interest system. The former to ensure that they fully understand the scope of work, the expectations being placed on them by BAUK, the requirement to fill in the inspection worksheets as the inspection progresses, and the item of interest system, as well as allowing them to ask any questions they may have.

The item of interest system also takes the pressure off the inspection team, as if they are not sure about something then they do not have to make the call as to whether its an anomalous or not, instead they fill in a form attached to the inspection scope of work and send it complete with photographs or video to the BAUK Marine Superintendent and the independent Marine TA, who then make the call as to which of the above containers it should be recorded in. As might be expected, the Inspection Note container is there to stop items being reported year on year that are not posing an integrity threat unless they become excessive.

The use of detailed inspection scopes of work, along with its associated containers, inspection worksheets and reporting methodologies has been widely welcomed by inspection teams, particularly as BAUK encourages their feedback and have incorporated several changes suggested by such teams, thus they feel they have part ownership of the scopes of work and so buy into them.

4. UNMANNED AERIAL VEHICLES

4.1 INTRODUCTION

Unmanned Aerial Vehicle (UAV), colloquially known as a drone, is the basic aerial vehicle. It can consist of a single propeller or jet driven unit, such as the Predator drone operated by a number of military forces worldwide, or a twin rotor helicopter type configuration such as the "baby shark" Schiebel Camcopter S-100 UAV currently being trialled by the UK HM Coastguard to provide safety overwatch and monitoring to its rescue helicopters, through to octocopter drones such as Intel© Falcon 8+ UAV as shown in Figure 4. Most UAVs though utilise four horizontal propellers and are either of an open configuration, such as the DJ Matrice 210 shown in Figure 5, or a caged configuration, such as the Elios 2 shown in Figure 6, with the main difference being that the latter is collision tolerant unlike the former but that is at the expense of flight time, payload capacity and wind limits for flying, due to the weight and drag of the cage.

The UAV will be equipped with a camera for the pilot, a flight stabilisation system, and may have additional systems such as collision avoidance, follow modes and an auto-pilot to allow autonomous flight, plus on board lighting. For survey work, a payload is typically needed consisting of an independent zoomable camera, preferable on both a pan and tilt gimble in order that it can be operated without interfering with the pilot's frame of reference. By adding such a payload, the UAV becomes a UAV system or UAVS, which is not to be confused with UAVs. This means that a UAVS requires a team of two personnel, a pilot / technician and a payload operator / inspector with team communication being crucial to the success of any inpection workscope.

BAUK has been making use of UAVS for survey work since 2018, with the first survey being of the above water hull exterior, accommodation block, helideck, freefall lifeboats, and flare tower. The survey work on the later included both visual spectrum and infrared spectrum imagery whilst the flare was burning, something that could not have been easily achieved except using a UAVS. The total survey time being two shifts for the hull et al and one shift for the flare tower, which is a considerable time, operational impact and cost saving over utilising rope access technicians offshore as well as a large reduction in the risk associated with the survey. In that instance, a DJ Matrice 210, as shown in Figure 5, was utilised as there was plenty of space for the UAVS to be operated, with take off and landing taking place from either the main deck or the helideck, dependent on the prevailing conditions and the portion of the vessel being surveyed at the time. This latter point is important, as whilst there are UAVs available that are driven by either petrol engines or jet engines, these are not particularly suitable for use offshore and certainly not when considering confined spaces. As such, the brushless motors of the UAV and its payload are operated via lithium polymer rechargeable batteries that have a finite amount of power available, which in turn means that there is a finite amount of flight time available before the battery packs need changing. Hence each flight needs to be carefully planned to maximise the useful survey time, whilst minimising transit time. This also means that battery management and recharging facilities becomes key to each, and every, UAVS survey.

4.2 LEGISLATIVE CONSIDERATIONS

Within the UKCS the use of UAVs is governed by the Air Navigation Order 2016 as amended [1], which is a statutory instrument under the Civil Aviation Act 1982 [2]. In practical terms, this means that the regulation of UAVs falls under the remit of the Civil Aviation Authority (CAA). The situation was made somewhat more complex by the implementation of the 2019 European Union Regulations with respect to the rules and procedures for the operation of unmanned aircraft, the full applicability of which was delayed until the 31st December 2020.

The result of the European Union Regulations, which the CAA consolidated into the Civil Aviation Publication (CAP) CAP1789A [3], is that practically all UAVs, except toy ones of under 250 grams, fall within the regulations. Whilst the regulations are somewhat drier than Classification society rules, the CAA have luckily provided a helpful guidance document in the form of CAP 722 entitled "Unmanned Aircraft System Operations in UK Airspace – Guidance" [4]. The guidance is not short, at some 238 pages but given it covers all UAVS from a Predator down to a Model Aircraft, which fall within the regulations, a lot of it does not apply to the type of UAVS that would be suitable for external or internal survey work offshore.

Due diligence is required though on the part of the company engaging the survey contractor to ensure that the contractor does comply with CAP 722 [4], which has a number basic and somewhat obvious threads:

- Legal Considerations;
- Airspace Usage & Management;
- Personnel Competence;
- UAVS Airworthiness, Flightworthiness & Certification;
- Safety Management including maintenance, and;
- Operational Requirements including flight planning & risk assessment.

Hence the use of Texo DSI by BAUK, given the former is solely an unmanned aerial vehicle survey company, and so is highly focused on meeting the CAA requirements, as well as customer expectations. This is shown by the CAA permissions held by Texo DSI as follows:

- Extended flight - up to 1,500 metres from pilot using extended visual line of sight (EVLOS) rules;
- Close inspection at height – fly over or within 10 metres of a structure above 400 feet;
- Close inspection low down – fly within 5 metres of any person, vehicle or structure below 400 feet, and;
- Crowd avoidance – fly within 50 metres of an organised, open-air assembly of more than 1,000 people.

The key to a successful partnership being early engagement, as discussed in more detail in Section 5.

4.3 CLASSIFICATION SOCIETY RULES & GUIDANCE

The International Association of Classification Societies (IACS) were at the forefront of allowing UAVS to be used in Class surveys by updating their Note 42 entitled "Guidelines for Use of Remote Inspection Techniques for surveys" [12] to include UAVS. ABS [5] and Lloyds Register [6], followed closely by ClassNK [7], were the first Classification societies to produce specific guidance notes on using unmanned aerial vehicles. Of these the ClassNK "Guidelines for Use of Remote Inspection Techniques for surveys" [7] gave a great overview whilst the "Guidance Notes on using Unmanned Aerial Vehicles" [5] by ABS provided a very practical approach to follow that included the wide considerations that come into play with remote inspection techniques such as data management, cyber-security, battery management and records.

In addition, shortly after the guidance was published from the aforementioned Classification societies, Oil & Gas UK (OGUK), which represent UKCS field operators and Duty Holders, published their own very comprehensive guidance entitled "Unmanned Aircraft Systems (UAS) Operations Management Standard and Guidelines" [8]. The OGUK guidance [8] covers not only survey requirements but also what needs to be in place to meet the requirements of CAP 722 et al [4][3]. Hence, in theory, if an operator follows the guidance in the OGUK guidance [8] then they should comply with their legal duties and be well on the way to having a successful UAVS inspection campaign.

4.4 SAFETY CONSIDERATIONS

All the guidance discussed in Section 4.3, with the exception of the IACS Note 42 [12], emphasises that a risk assessment should for the basis, with ABS stating:

"A case-specific risk assessment should be carried out to identify any hazards related to any planned remote operations and the need for risk control measures. It is recommended that this risk assessment be finalized during the survey planning meeting and incorporated in the Survey Planning Document. Each party should acknowledge the risks associated with the remote inspection activities and agree to the mitigation plan associated with those risks."

It then provides the following guidance that the risk assessment should include, but also not be limited to:

- Explosion Risks in Hazardous Areas, including the payload, battery and operations;
- Dropped Object Risk in the event the UAV fails or malfunctions;
- Collision risk, particularly where visual line of sight (VLOS) is not maintained or upon unexpected interruption of pilot operation e.g. someone speaks to the pilot and so distracts them;
- Lost Link Risks associated with the loss of the communication control link to the UAV, particularly if operated in an unreliable radio frequency (RF) environment or if nearby systems interfere with the UAV's RF;
- Other risks to the operator, UAV or asset such as the working environment, ongoing operations near the UAV survey and emergency or operational scenarios that may or may not include the UAV but that could impact its operations, such as flaring during a flare tip inspection.

BAUK with the assistance of Texo DSI and ACL undertook such a risk assessment for inspection a COT from the marine deck of the Armada Kraken FPSO with the default position being that there would be not tank entry involved. This assessment covered a number of areas but of most relevant from a safety perspective were:

- Spark Potential Risks;
- Collision Risks;
- Lost Link Risk;
- Power Management Risks, and;
- Risks associated with retrieval of a downed UAV.

Whilst from an operational perspective the following areas were of concern:

- Navigation;
- Battery Management, and;
- Transportation Considerations.

The main issues associated with each being discussed in the following section along with the solutions that were adopted to ensure that the safety and operational risk was as low as is reasonably practicable (ALARP), which is the basic test for all UKCS risks.

5. RISK MANAGEMENT

5.1 SPARK POTENTIAL RISK

As discussed in Section 4.1, for offshore inspection work only battery powered UAVs are considered suitable for offshore inspection work at this stage. Given that these make use of brushless motors for both flight and camera gimbals then in flight their spark potential is extremely low. Thus, the main spark potential comes when changing a battery pack and having to either un-plug or plug in a pack given their current producing capacity. Given that batteries are never changed out with the UAVS switched on then the procedures surrounding changing battery packs by default minimise the risk potential. Hence, by operating under a hot work 'spark potential' permit with the requirement for a local gas detector, then the risk of a UAV causing any sort of fire or explosion is minimal. The larger risk is actually from line-of-sight fire detectors 'seeing' any potential flash when the low voltage battery pack is connected and so tripping the production plant and causing a system wide blow-down and flare off, which inherently has its own safety, operational and commercial risks. To overcome this, a weather shelter was built on the marine deck for the survey team to operate from and change the battery packs within, which provided a wall of tarpaulin in the direction of the line-of-sight fire detectors. It also provided a measure of weather protection to the survey team as well as keeping glare off the survey team's screens, all three of them. One for the pilot, one for the payload operator / inspector and a separate one for the Verification / Class surveyor so that they did not need to look over the shoulder of another team member and so risk distracting them.

5.2 COLLISION RISK

With the pilot having to fly the UAV in a dark fully enclosed space without actually being able to see the UAV once the pilot has flown through a man-hole not that much bigger than the UAV, there was a high probability that the pilot, no matter how skilled, would fly into an obstruction within the COT. The difficulty of flying a UAV using an onboard camera plus sensors should not be underestimated, hence the use by BAUK of a specialist UAV survey company in the form of Texo DSI.

Given that contact between the UAV and the ship structure inside the tank is almost inevitable then an open frame UAV, such as that shown in Figures 4 & 5, is not viable as contact would mean a damaged propeller and so a potentially crashed UAV. Hence the only choice to be made comes down to what type of caged UAV to use, which in this instance was a Elios 2 as shown in Figure 6. It had originally been the plan to use an Elios 1, but a delay to the scope implementation meant that the Elios 2 was available. The delay working in BAUK's favour as unlike the Elios 1 the cage on the Elios 2 does not rotate on impact, potentially getting in the way of the video camera, instead it elastically deforms. The Elios 2 also has improved flight controls and visual 'follow' modes for tracking along girders and stiffeners.

In terms of collisions with personnel, the area was cordoned off around the weather shelter and tank entrance so that the survey team had a clear area to work in.

As might be imagined the survey created a lot of interest in the crew who were keen to see the UAVS in action, which the survey team managed in a controlled manner. This, as might be expected, satisfied the crew's natural curiosity and helped them 'buy into' the survey and the use of its results in safeguarding the integrity of the asset.

5.3 LOST LINK RISK

One of the biggest unknowns before the offshore phase of the project was how well the pilot and payload operator would be able to control the UAVS from outside the COT given that the COT itself would act like a Faraday cage. It was therefore decided early on that a range extender unit would be lowered part way into the COT through one of the portable Crude Oil Washing (COW) machine ports. A range extender is in effect an antenna with a signal booster built into it on the end a length of cable that is plugged into the pilot's control unit instead of its normal aerial which improves the remote to UAV control signal and signal range. These details were included with the Project Specific Procedures produced and refined through testing by Texo DSI. Even with these measures in place, there was always the risk of a loss of communication and so loss of control of the UAV necessitating its retrieval from the tank either after a crash or a deliberate landing.

5.4 POWER MANAGEMENT RISKS

Other than communication issues or collision, the other main reason that a UAV may of crash or have to be deliberately landed within a tank is due to running out of battery power. Given the risks that would be involved with retrieving a 'downed' UAV then power management was given high precedence within the Texo DSI project specific procedures. One of the problems with running on battery packs is the lost survey time due to the time it takes to transit from the survey site to the operator, change the battery pack and then transit back to the survey site. Hence from a survey efficiency perspective, in the main one would want to maximise the flight time on each battery pack, however, with the possibility of having difficulty finding the exit when flying blind combined with the retrieval risk then each individual flight was curtailed when the battery reached 50% capacity such that there was plenty of power length for existing the COT safely.

5.5 RETRIEVAL

One of the biggest drivers for undertaking UAV COT surveys is to reduce the amount of time (or remove) that personnel need to spend in such tanks, thus mitigating the risks to personnel on board the asset. In addition, whilst entering a cargo tank under double block and bleed isolation is fairly typical marine practice worldwide, it is not it is not seen as a positive form of isolation on offshore oil & gas assets operating within the UKCS. To achieve UKCS standards, the first thing the inspection team would need to do is to swing the blinds on the cargo lines whilst wearing and using breathing apparatus. Whilst this is not impossible, due to the way the MT Prisco Alcor was originally setup, it would certainly be challenging for the inspection team to do, hence, there is a desire to minimise as far as is reasonably practicable the need to enter any of the cargo tanks.

Obviously it would not be practical to undertake a UAV based survey if it would not be possible to enter the tank to retrieve a downed UAV, so after careful risk assessment it was decided that in the event of an issue with the UAV that prevented it from being flown to outside the tank, it should be landed on the bottom of the tank if at all possible to make it easy to retrieve, and then a confined space entry team would retrieve it whilst wearing and using breathing apparatus. The breathing apparatus being necessary despite the 'gas free' nature of the tank to mitigate the risks from fumes 'passing' the isolation valves.

As with all manned tank entries, the ability to be able to undertake it depends on the state of the walkways, stairs and ladders from a safety perspective. Hence the decision-making protocol for whether to carry on with the UAV survey after the orientation flight, as shown in Figure 7, was developed, which is based on initially surveying the accessways and then dependent on their condition making a decision as to whether the survey could continue as planned or whether a rope access team needed to be on board to retrieve UAV if necessary or whether survey needed to be curtailed.

From Figure 7 it can also be seen that it was intended to use a tether attached to the drone during the first flight to allow its retrieval from out with the tank. Once on-site, the survey team rapidly found that there were so many potential snagging points for the tether, that using one could well have caused the UAV to become snagged and so require retrieval, rather than reducing the retrieval risk. Hence, the survey team went through a management of change process with the offshore personnel, complete with risk assessment that assessed the risks associated with using the tether versus a controlled build-up of the length of each non-tethered flight as the survey team became familiar with the tank layout, which resulted in the latter being the method adopted. Consequently, the decision protocol, shown in Figure 7, will be updated to reflect that before the next survey, this being one of several lessons learnt during this successful campaign.

5.6 NAVIGATION

Given that the pilot was flying the UAV from outside the tank without any special lighting or other visual aids, then it can be readily understood that navigation within the tank was always going to be something of a challenge, including just finding the exit at the end of a flight. This is where the surveyor / payload operator has a big part to play, as one of their jobs was to always keep track of the position of the UAV, which is essential from a survey perspective anyway, but absolutely essential when transiting to or from the worksite within the COT.

One thing that did aid the team with orientation was having all the portable COW ports open as from the underside of the deckhead they create a distinct light pattern that can be used to globally locate the UAV within the COT simply by looking up.

The teamwork between the pilot and surveyor / payload operator was honed by Texo DSI by bringing the team together well before the offshore phase, such that they could trial blind flying within the company's warehouse, both with the lights on and off. By doing this, Texo DSI's team were not only able to work out the best way to operate as a team for this specific type of survey, but also to then incorporate the lessons learnt into project specific inspection procedures. These procedures were then tested and refined prior to the offshore phase, thus when the team arrived offshore they knew exactly what they needed to do and how they were going to do it.

From these trials it became obvious that intense concentration was going to be required during each flight and so unlike with other surveys it would not be possible for the Verification / Class surveyor to just ask the pilot to stop and have a look at something, instead a protocol was needed within the procedures for dealing with such eventualities that satisfied all parties. The protocol was extremely simple, if the Verification / Class surveyor saw something on their dedicated screen that they wanted a closer look at they were to note it down, then before the start of the next flight the UAV team would ask if there was anything from the previous flight the surveyor wanted to have another look at. If there was, then this became the priority for the next flight before carrying on with planned inspection tasks, and in practice this system worked very well, but as with all such things it needs to be discussed and emphasised during the pre-mobilisation briefing, such that everybody is singing from the same song sheet.

DNVGL who are the are Independent Verification Body (IVB) and Classification society for the Armada Kraken FPSO were involved in the plan to use UAVs for COT surveys from the point of conception onwards. Hence their views, suggestions and requirements were taken on board by all parties through the planning and execution phases, through to lessons learnt. One of the outcomes of this early engagement was Texo DSI becoming one of the first UAV survey companies to be certified for "Survey using remote inspection techniques (RIT) as an alternative means for close-up survey of the structure of ships and mobile offshore units" by DNVGL under their Approval of Service Suppliers Programme (DNVGL-CP-0484) [9], which is based on the based on IACS unified rules UR Z17 [10] and so is applicable to other Class societies.

5.7 BATTERY MANAGEMENT

As discussed in Section 5.4, all flights were curtailed and the UAV brought out of the tank when the remaining charge within the battery pack was showing on the pilot's screen as being down to 50%. This meant that each flight was only in the region of six to eight minutes long and so during a 12-hour shift the UAV team got through up to twenty battery packs. These not only have to be provided and transported, which has its own issues as discussed in Section 5.8, but also need to be stored ready for use and recharged either during the shift or before the next shift.

As can be imagined, this creates something if a logistical headache as it is simply not practical to have one charger per battery pack. Consequently, in agreement with the offshore crew the team were able to make use of a desk within the Forward Temporary Refuge that was equipped with a number of sockets, which enabled them to charge four battery packs at a time. Thus, by changing the battery packs over in the chargers every four flights or so, they were able to stay on top of the battery charging requirements.

It is all well good having a multitude of batteries charged up, but their storage before use, plus the mass charging of battery packs and transporting battery packs to the worksite poses its own risks that need to be accounted for, given that a cracked lithium polymer battery cell can suffer from a thermal runaway event, which can also occur if it gets too hot or is over-charged. Consideration therefore has to be given to ensuring that the correct fire extinguisher medium is available, and the emergency response teams are aware of the presence of the batteries. Thermal runaway events should not be under appreciated given that lithium polymer battery fire can be self-sustaining as the decomposition process produces oxygen. Although this will be covered under the Safety Management System's Permit to Work on a UKCS offshore installation, consideration also needs to be given to it at the planning stage given that such events can lead to flame temperatures of 800° centigrade according to Dongxu et al [11].

In this instance, to minimise the risks associated with storing and transporting the battery packs on the asset, the battery packs were kept in their dedicated transportation case at the worksite unless they were either in use or being charged. Whilst the battery packs being taken to or from the charging site were carried in Lithium Polymer (LIPO) bags, which are glass-fibre lined and so protect the operator and the asset from a battery pack suffering a thermal event. It might seem counterintuitive to store the batteries on the marine deck, albeit in their storage case, however, the hull structure and onboard fire detection and suppression systems are designed, as well as the emergency response teams are trained, for dealing with crude oil pool fires on the marine deck. Given that a crude oil pool fire can reach temperatures in excess of 1,400 centigrade, then the fire suppression system and ship structure can cope with the lower temperature of a lithium polymer batter pack thermal runaway using either high volume water deluge to cool and disrupt the thermal runaway or foam smothering, which will also cool and disrupt such an event.

5.8 TRANSPORTATION CONSIDERATIONS

Lithium polymer battery packs are classed as dangerous goods under the International Maritime Dangerous Goods (IMDG) Code [13], complete with their own separate amendment 38-16 giving rise to Special Provision 188. In this context though, it means that the battery packs and their cases need to comply with the code, only a limited quantity of battery packs can be sent offshore by boat at any one time, and extra time is required for shipping due to the freight forwarding firm undertaking the necessary checks. This process when moving such goods offshore is never an easy one and, in the past, has caused items that an UAV or ROV team might use routinely, such as isopropanol in aerosol form, being refused by the freight forwarder, despite the fact it can be bought at my auto-parts stores.
Early engagement with the freight forwarder is therefore essential to prevent project delays and so lost survey time.

5. WORKFORCE ENGAGEMENT

Although BAUK 'owned' the project, it was from the conception stage onwards very much a tripartite project between BAUK, Texo DSI and ACL, with each party being responsible for discrete parts of the overall scope, namely:

- BAUK: engagement with onshore and offshore workforces to get their 'buy-in' to the project, project risk assessment, logistics, project document control, liaising with DNVGL, and co-ordination between the various parties.
- Texo DSI: provision of UAVS and survey team, proof of concept, development of project specific procedures including testing said procedures and refinement thereof.
- ACL: development of work-scope, reviewing and approving project specific procedures on behalf of BAUK, answering queries to survey scope and providing input it during survey as necessary, such as reviewing Items of Interest and providing guidance to survey team.

5.1 BUMI ARMADA

5.1 (a) Onshore

The first task after the onshore asset integrity team had decided to investigate the use of UAVs for COT inspections, was to engage with the management team that this was a viable survey methodology and had the potential to:

- Shorten the time an individual COT was out of service due to being surveyed;
- Significantly reduce the cost of surveying each COT;
- Provide almost the same level of integrity assurance as a manned entry would when using scaffolding or rope access techniques, if not in some respects higher;
- Significantly reduce the risk to personnel undertaking the survey, and;
- Significantly reduce the number of personnel that needed to be dedicated to the survey, which also frees up beds on board for other vendors.

Given the above points the BAUK management gave their full support to the COT UAV surveys.

5.1 (b) Offshore

Due to the way in which the UKCS offshore safety legislation is setup, the ultimate control of what happens on an offshore asset lies with the Offshore Installation Manager (OIM) and the management team on the asset that answer to the OIM. Hence the next stage was for the onshore integrity team to engage with the offshore management team to explain the intended scope, how it was intended to be carried out and the measures already taken, such as the 'proof of concept' flights undertaken by Texo DSI and witnessed by BAUK and ACL personnel. Once the offshore team had 'bought' into the concept then the onshore integrity team arranged a risk assessment chaired by the BAUK HSE manager to cover the whole of the intended scope, including tank entry to retrieve a 'downed' UAV under double block and bleed isolation.

The risk assessment gave the offshore team a forum in which to discuss any concerns, and to hear what mitigations were already planned, as well as discussing other potential mitigations, and what the offshore team needed to do to prepare for the survey. This preparation amounted to the following:

- Based on the present loading rate calculating how long a pair of COTs could be kept out of service whilst keeping the hull within limits;
- Verifying the isolation sequence for the tanks;
- Preparing Permits to Work for the UAV survey and for tank entry to retrieve a 'downed' UAV, including the necessary offshore mitigations agreed in the risk assessment, so that the permits were 'ready to go' once the survey team arrived offshore, and;
- Undertaking a trial isolation, hot water wash of the COT and gas freeing, such that any procedural changes required could be executed before the survey team arrived. As well as bringing the COT back into service after its isolation.

5.2 TEXO DSI

As discussed earlier, Texo DSI were engaged early in the project, which greatly assisted its overall execution, as it now only allowed the BAUK and ACL project team to fully understand the capabilities of the UAVS and see it in action in a purpose-built test tower, but to ensure that the project was set up in a UAVS friendly manner without requirements that would be impossible for Texo DSI to fulfil. To facilitate this, Texo DSI nominated the main body of their survey team early on, such that they could see the project through from start to finish. The early supply of the ACL developed inspection work-scope to the Texo DSI project team by BAUK also helped to 'educate' the team as to the BAUK specific requirements. As with all inspection teams to date, the Texo DSI survey team welcomed having such a detailed and clear scope of work document, as they then knew specifically what BAUK required and could see the benefit of the containers system, which takes the pressure off them when it comes to making judgement calls on anomalies.

5.3 DNVGL

Discussions with DNVGL as both the IVB and Class were had early on in the project to ensure that the UAV survey would be credited for both verification activities and Class. At project outset in 2018, it was not clear whether this would be the case due to DNVGL having its own UAV unit and seemingly only crediting surveys undertaken by them. As the project progressed though, DNVGL ratified IACS UR Z17 [10] and updated their Approval of Service Suppliers Programme rules [9], thus enabling Texo DSI to become an approved supplier, which meant that if the surveys were carried out to the satisfaction of the attending DNVGL surveyor then they would be credited for both IVB and Class.

In addition, for the last decade the personnel involved from BAUK, DNVGL, and ACL have been working together on a number of assets. Hence, the methodology being adopted for the COT survey in terms of a detailed inspection work-scope and the use of containers to simplify the survey and its findings, was familiar to the DNVGL focal point surveyor and they are supportive of it, as it makes their life easier too.

6. SURVEY EXECUTION

6.1 PREPARATION

As discussed in Section 5.1(b), prior to the mobilisation of the UAV survey team and IVB/Class surveyor, the offshore team undertook a full 'dress rehearsal', where they isolated the subject COT, hot water washed it, gas freed it and then opened it up ready for inspection. In this instance though the inspection consisted of shining a torch into the COT to look at how clean the hot water wash had been able to get the tank. As the Armada Kraken FPSO produces heavy crude, this has to be kept heated in the tanks to prevent it from solidifying, with the net result that water in an adjacent tank ends up becoming pre-heated and from there is can be directed through the COW heater before being used to wash the tank. This was found to work well and was repeated prior to the survey team arriving on site, such that the tank was ready for the inspection as soon as the team had arrived on board.

Although the temporary structure at the worksite could have been put up before the team arrived, and probably next time it will be, it was felt in this instance that given how little time it takes to erect such a shelter that it worth waiting until the survey team arrived on site to ensure that it met their requirements.

6.2 PRE-MOBILISATION BRIEFING

BAUK always thoroughly brief inspection teams before mobilisation to ensure that the inspection is carried out, as far as is reasonably practicable, as per the detailed scope of work, which includes going through the container system even for teams that have previously deployed. It also allows the survey team to ask any last-minute questions about the work-scope that may have arisen.

The other purpose of the pre-mobilisation briefing is to ensure that the attending IVB/Class surveyor, where it's not the focal point surveyor, understands correctly the intent of the survey, as it may be in excess of their requirements, the containers system for potential anomalies and how they fit into that, and any specific instructions there may be with regard to survey execution. In this instance, this latter point covered not speaking to the pilot or payload controller during a flight, as they needed all their concentration to operate the UAVs from outside the tank and navigate their way around, as discussed earlier. It also covered the fact that there would be a separate screen for the surveyor and the opportunity to review the footage at the end of each shift if they so desired.

Due to the COVID-19 situation the pre-mobilisation briefing had to take place remotely, however, BAUK and ACL have undertaken remote briefings at times prior to COVID-19, so switching over to that did not present any issues.

6.3 WORKSCOPE EXECUTION

As per the agreed plan, the Marine Supervisor offshore had the tank hot water washed and gas free before the four-man Texo DSI and DNVGL surveyor arrived on the asset. This meant that after the team had undertaken their arrival briefing, they could visit the worksite and get everything set up including advising the scaffold team as to the requirements for the temporary weather shelter.

Once the initial se-up was complete, which included sorting out where to site the battery charging facilities, the Texo DSI team were able to carry out orientation flights that evening before moving on to start the survey proper. Over the four shifts it took to complete the survey of the COT tank, which included every frame and stiffeners, a job made easier by the automatic follow mode the Elios 2 drone has, the following flights were carried out to the satisfaction of DNVGL:

- 06/07/20 (NS) 7 flights were carried out
- 07/07/20 (DS) 15 flights were carried out
- 07/07/20 (NS) 15 flights were carried out
- 08/07/20 (DS) 20 flights were carried out

Thus in two days the UAV survey team had managed to accomplish what it would have taken two weeks to achieve through the use of scaffolding and probably longer if rope access technicians had been used. The quality of the survey, which was filmed in 4k Ultra High Definition, which is cinema grade, was aided by the unique onboard anti-static lighting on the Elios 2, which minimises the impact of airborne dust on the quality of the visual image.

7. CONCLUSIONS

Through the 6th to the 8th July 2020 Texo DSI acting on behalf of Bumi Armada UK Limited successfully completed what is believed to have been the first UAV based inspection of a Crude Oil Tank from completely outside the tank, with no manned entry, with the Armada Kraken FPSO still producing crude oil and filling adjacent tanks. This was made possible by the close cooperation between BAUK, Texo DSI, A Comley Limited and DNVGL, with credit to the survey team and the BAUK offshore team for the professional way the survey was undertaken.

The main lessons that were learnt from the survey were:

- A Detailed Scope of Work is well worth developing for all inspections, particularly those that involve remote inspection techniques, as they ensure that all parties know what is required, where that can not be achieved then the acceptable level is agreed, and the scope updated accordingly.

- The use of anomaly, item of interest and notes containers for items spotted by the survey team, complete with a simple formal review process, takes the pressure off the survey team with regard to making judgment calls over the integrity status of items, as well as improving the integrity of the asset as the onshore integrity team then has a better idea as to the actual status of the asset.

- A detailed pre-mobilisation briefing of the project team and Class surveyor ensures that everyone knows what is expected of them, knows how the container system works and how and when to apply it, including who to contact for advice and how.

- Early engagement with BAUK personnel, onshore and offshore, and the Texo DSI project team ensured the success of the project by ensuring that everybody 'bought into' the objectives and key trials were carried out ahead of time, such that when the mobilisation happened everything was already proven and ready to go.

- Early engagement with DNVGL enabled Texo DSI to become accredited as a service supplier so the inspection has been Verification & Class credited.

- Development by Texo of project specific procedures based on onshore trials meant that their survey team were fully prepared for the task that they went out to the asset to do and had worked out the best way to work as a pilot and payload operator team to complete the mission efficiently and effectively.

- A suitable level of tank cleanliness can be achieved by using hot water washing alone, without the need for crude oil washing, thus saving time, effort and making the job easier to complete particularly given the type of crude oil that the asset produces.

- The suction wells could not be properly inspected by the UAV as there simply was not enough access, however, the underside of them is accessible from the water ballast tanks and so they can be inspected visually and by using Ultrasound.

8. ACKNOWLEDGEMENTS

The authors wish to thank Bumi Armada UK Limited, Bumi Armada Berhad, Texo DSI and A Comley Limited for the opportunity to write this paper and disseminate the lessons learnt by the authors through the project.

9. REFERENCES

1. HMG, '*Air Navigation Order 2016 as amended*', 2016/765, August 2016
2. HMG, '*Civil Aviation Act 1982*', 1982
3. Civil Aviation Authority, '*The UAS Implementing Regulation*', CAP1789A, January 2021
4. Civil Aviation Authority, '*Unmanned Aircraft System Operations in UK Airspace – Guidance*', CAP 722, November 2020
5. ABS, '*Guidance Notes on using Unmanned Aerial Vehicles*', November 2016
6. Lloyds Register, '*Guidance Notes for Inspection using Unmanned Aircraft Systems*', March 2016
7. ClassNK, '*Guidelines for Use of Remote Inspection Techniques for surveys*', April 2018
8. Oil & Gas UK, '*Unmanned Aircraft Systems (UAS) Operations Management Standard and Guidelines*', January 2017
9. DNVGL, '*Approval of Service Suppliers Programme*', DNVGL-CP-0484, February 2019
10. International Association of Classification Societies, '*Approval of Service Suppliers Programme*', IACS UR Z17, October 2020
11. DONGXU, O., MINGYI, C., QUE, H., JINGWEN, W., ZHI, W., JIAN, W. 'A Review on the Thermal Hazards of the Lithium polymer Battery and the Corresponding Countermeasures', *MDPI Applied Sciences Journal 2019-9-2483*, June 2019
12. International Association of Classification Societies, '*Guidelines for Use of Remote Inspection Techniques for surveys*', IACS REC 42, June 2016
13. International Maritime Organisation 'The International Maritime Dangerous Goods (IMDG) Code', IMDG, 2018

9. AUTHORS BIOGRAPHY

Campbell Ross holds the current position of Marine Superintendent at Bumi Armada UK Ltd.

Campbell Ross is a Marine Superintendent with 33 years of experience in the Merchant Shipping and Oil & Gas Industries. Experience gained in various capacities on various vessel types including, general cargo, product tankers, passenger, roll on–roll of ferry, crude oil tankers and FPSO's.

Whilst employed as a Marine Superintendent, Campbells responsibilities included acting as focal point for marine operations and responsible for ensuring the appropriate support is provided to the Operations Team onshore and offshore.

He has co-ordinated with Technical Authorities and Inspection Service Suppliers in completing Hull and Mooring Inspections in accordance with Verification, Classification Society and Risk Based Inspection Plan requirements. Hull Inspections include COTs by way of UAV and WBTs by way of ROV.

Steve Moir holds the current position of Operations Director for Texo Asset Integrity. He is responsible for the safety, operational and financial performance of the organisation and provides strategic guidance advice and guidance to the Group board. Develops and maintains, cost effective, R&D programmes to ensure that the company remains at the forefront in the industry.

Steve is a Chartered Manager, with over twenty-five years' experience in the upstream oil and gas and marine industries.

He has led the introduction of multiple integrity technologies to the petrochemical and marine industries, developing the testing regimes to create the technical tools and competency requirements to deliver a credible, reliable and consistent products to the marketplace, including;

- Ultra-sensitive, metal-loss electrical resistance technology.
- A magnetic flux NDT screening technique to rapidly detect localised corrosion pitting to process piping.
- Numerous accredited visual inspection techniques, associated with Unmanned Aerial Vehicle techniques.

Andrew Comley is the Managing Director and Principle Naval Architect of A Comley Limited, which is a micro consultancy and engineering company based in the North East of Scotland. He is responsible for the day to day running of the company and its workshop, as well as providing expert consultancy, design engineering, fabrication and machining services to the company's clients.

He is a Chartered Naval Architect who has worked for the Advantica (now part of the DNVGL group safety division), the UK Health & Safety Executive, Noble Denton and Kingfisher Marine Services Limited, part of the Lithgows Group, prior to establishing A Comley Limited in 2001. During this time, he has been responsible for providing Technical Authority services to a number of clients, developing and writing detailed inspection workscopes for a variety of remote inspection techniques, and providing general advice and consultancy on a wide range of topics from mooring integrity to ROV tooling.

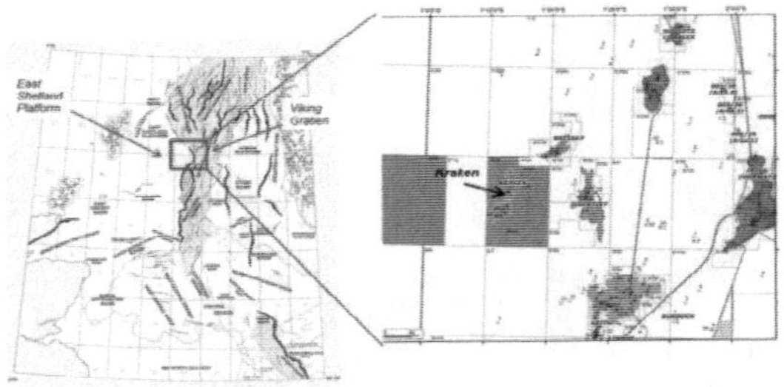

Figure 1: Aramda Kraken FPSO location

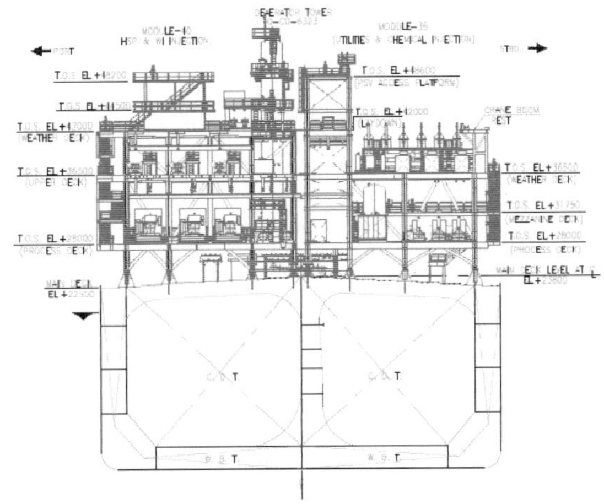

Figure 3: Cross-section at "A-A" looking forward

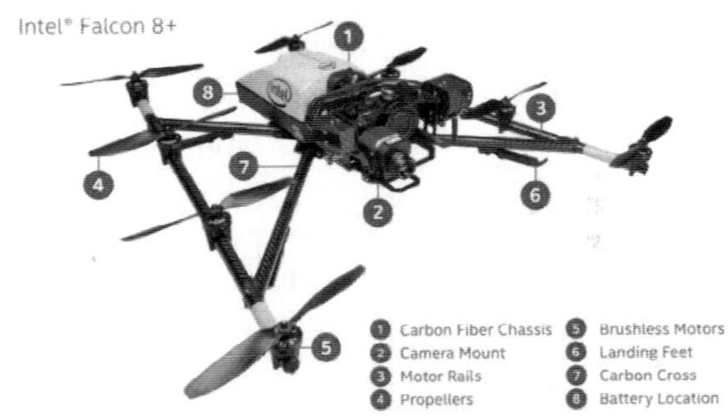

Figure 4: Intel Falcon 8+ (www.intel.co.uk)

Figure 5: DJ Matrice 210 RTK V2 (www.dji.com)

Figure 2: Armada Kraken FPSO Profile

Figure 6: Elios 2 Unmanned Aerial Vehicle System

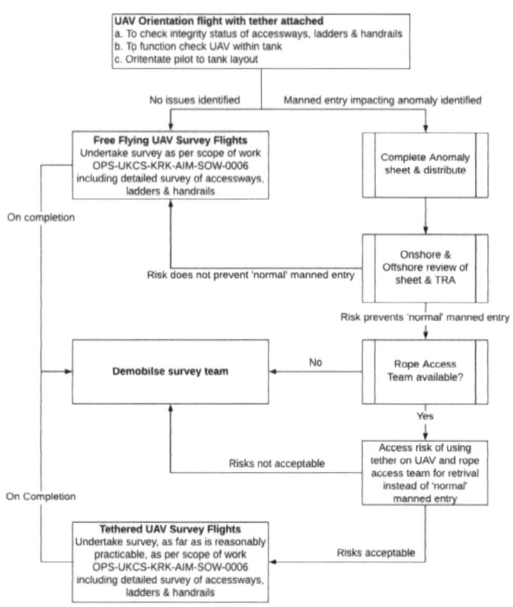

Figure 7: Crude Oil Tank Inspection UAV Decision making Protocol

Figure 8: UAVS about to descend into COT 1 Port on the Armada Kraken FPSO

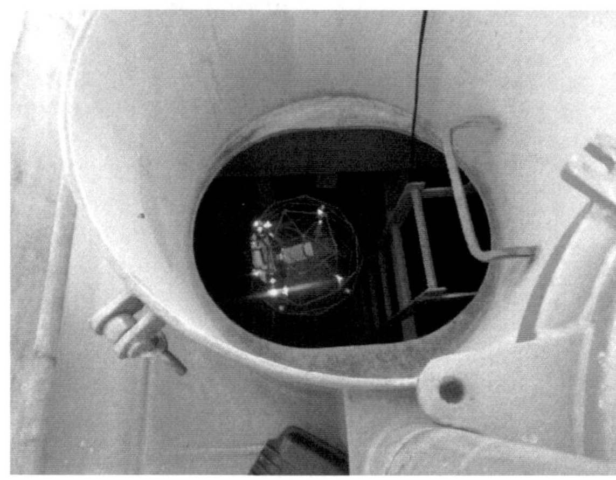

Figure 9: UAVS ends its descent into COT 1 Port on the Armada Kraken FPSO before having to transit away from the entry platform to access the main body of the tank

Figure 10: The last view of the UAVS the survey team get before having to control the UAVS purely from its onboard camera and sensors in the darkness of COT 1 Port on the Armada Kraken FPSO

The Royal Institution of Naval Architects
8-9 Northumberland Street
London, WC2N 5DA
United Kingdom

ISBN 978-1-7138-8304-3

CURRAN ASSOCIATES INC.
proceedings.com

International Webinar on Ships' Life-Cycle 2021

Online
31 March 2021

The Royal Institution of
Naval Architects